# Atoms and
# MOLECULES MEET

### Rebecca Woodbury, Ph.D., M.Ed.

**Gravitas Publications Inc.**

# Atoms and
# MOLECULES MEET

Illustrations: Janet Moneymaker
Design/Editing: Marjie Bassler, Rebecca Woodbury, Ph.D., M.Ed.

Atoms and Molecules Meet
ISBN 978-1-950415-11-3

Published by Gravitas Publications Inc.
Imprint: Real Science-4-Kids
www.gravitaspublications.com
www.realscience4kids.com

Image credits: Cover & Title page: alexeevich, Adobe Stock. Above: Nikolai Titov, Adobe Stock. p. 3: Monkey Business, Adobe Stock. p.11: Aleksey, Adobe Stock. p. 13, Alexandre, Adobe Stock. p. 15, Ajamal, Adobe Stock. p. 17, Natalieime, iStock. p. 19. simonidad, Adobe Stock. p. 21 fizkes ,Adobe Stock

RS4K

When you and your friends meet,
kids in the group might break apart
into smaller groups or switch partners
or rearrange into new groups.

The same thing happens when atoms and molecules meet.

Molecules can break apart, switch partners, or rearrange their atoms to make new molecules.

Atoms and molecules are making new friends!

Hi! Nice to meet you.

Water molecule

I am hydrogen.

Hydrogen

Howdy! We are a salt molecule.

Sodium chloride
(table salt molecule)

I am nitrogen. Nice to meet you!

Nitrogen

# Review

- **Atoms** are tiny building blocks that can link together.

- **Atoms** make everything we see, touch, taste, and smell.

- We can draw **atoms** with arms and hands to represent **linking electrons**.

- **Atoms** link together to make **molecules** using their **linking electrons**.

We are a
salt molecule.

I am an oxygen atom.
I have two arms.
I make two bonds.

**Salt molecule**

**Oxygen**

Hi! I am carbon.

**Carbon**

We are two atoms.
We can make one bond.

**Sodium**

**Chlorine**

When atoms and molecules meet and rearrange, it is called a **chemical reaction**.

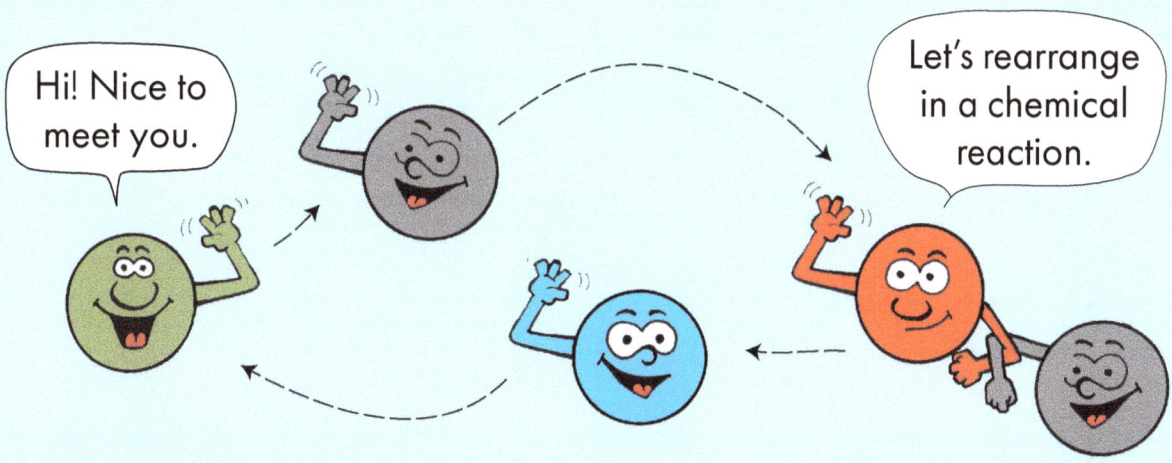

Sodium    Chlorine

Hydrogen    Oxygen    Hydrogen

You can see **chemical reactions** when they happen.

- You might see **bubbles**.

- You might see **fire**.

- You might see a **color change**.

- You might see small **particles**.

- You might see **smoke** or **fumes**.

You can see a chemical reaction when sodium metal is added to water.

The water breaks apart and rearranges to make **hydrogen gas** that burns as fire.

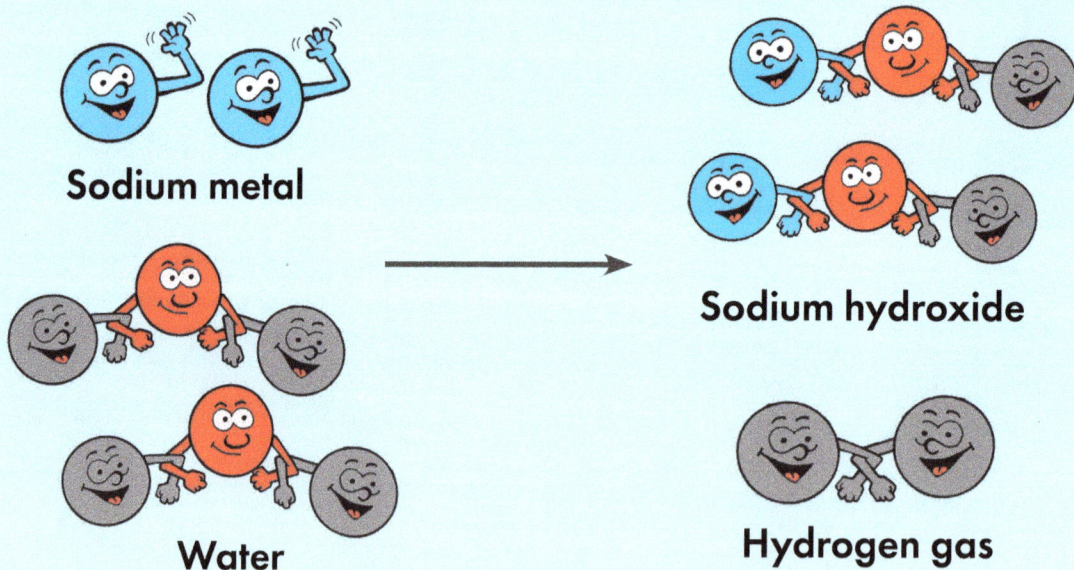

Sodium metal

Water

Sodium hydroxide

Hydrogen gas

Chemical reaction between sodium metal and water

You can see a **chemical reaction** when water breaks apart to make **hydrogen gas** and **oxygen gas** using **electrical energy**.

Bubbles of hydrogen and bubbles of oxygen fill the test tubes.

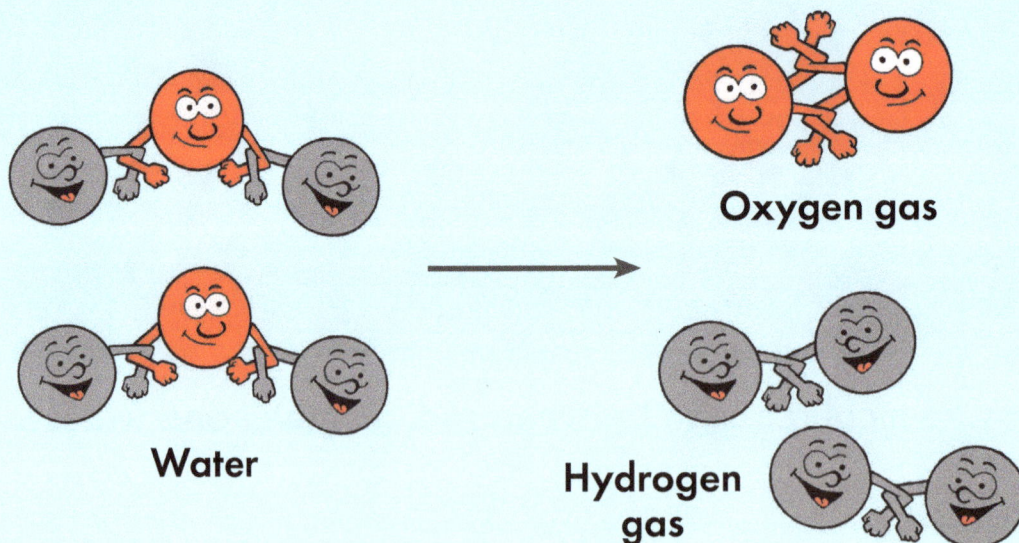

**Oxygen gas**

**Water**

**Hydrogen gas**

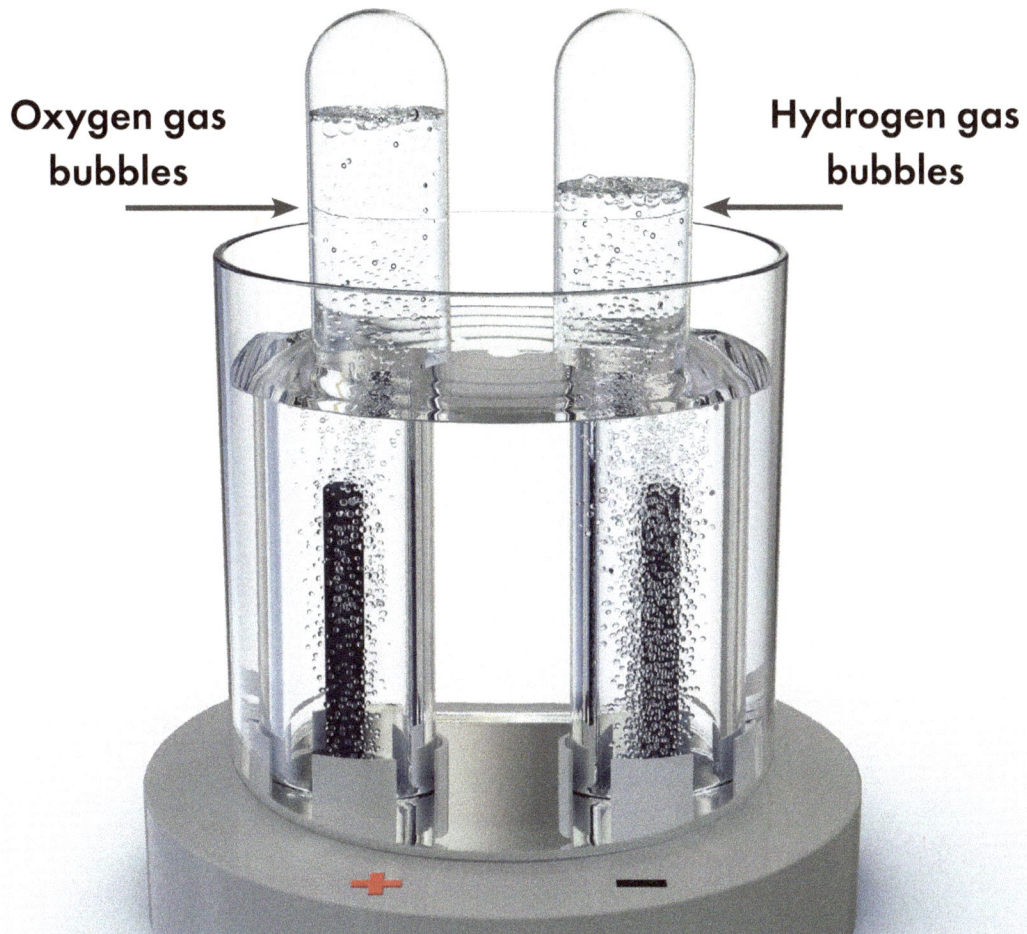

Oxygen gas bubbles

Hydrogen gas bubbles

Water can break apart to make hydrogen and oxygen gas.

You can see a chemical reaction when molecules made of lead, nitrogen, oxygen, iodine, and potassium rearrange in a chemical reaction to make yellow particles called **lead iodide**.

Iodine　　Lead　　Iodine

**Lead iodide molecule**

Pretty!

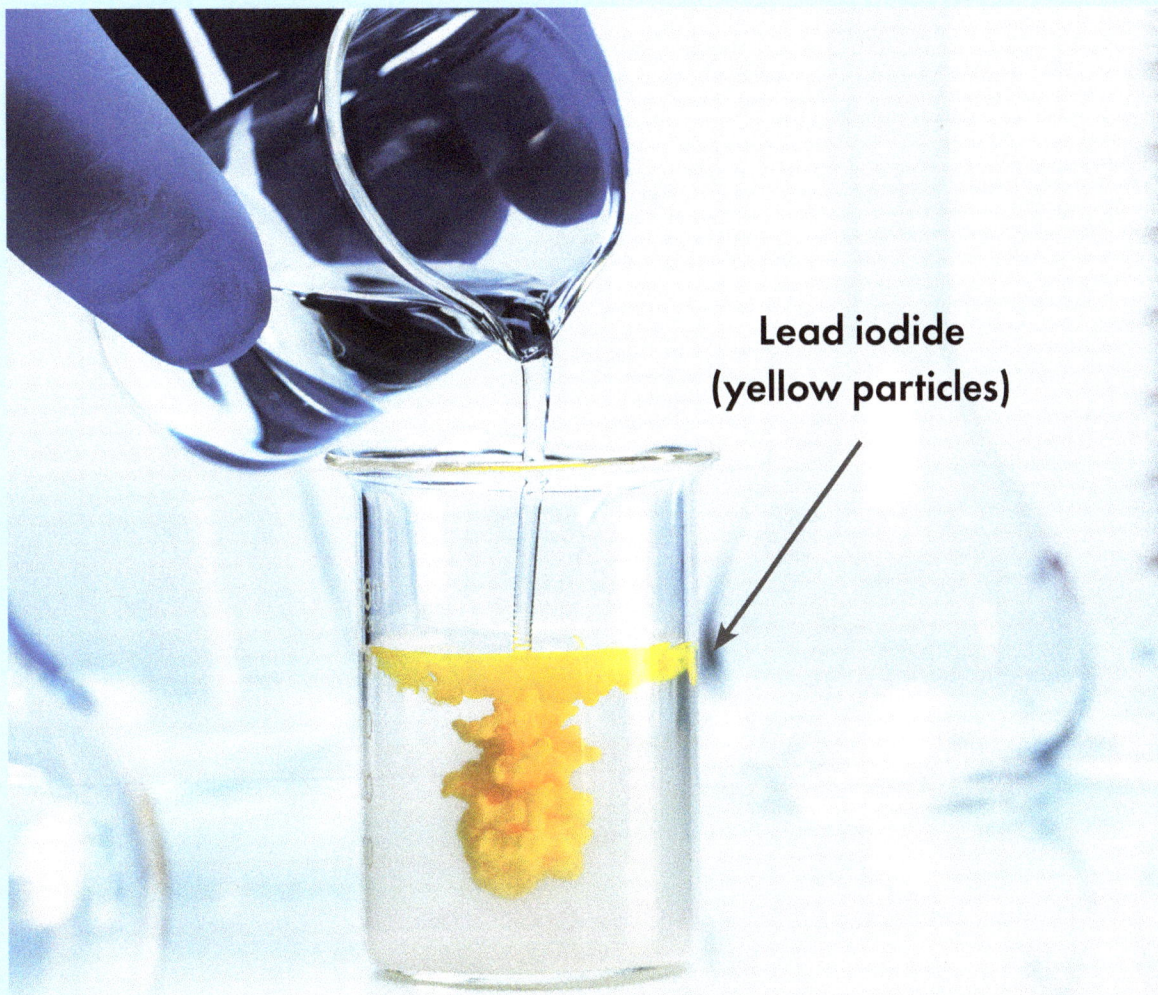

Lead iodide
(yellow particles)

Lead and iodine combine to make yellow lead iodide particles.

You can see chemical reactions when you make bread.

Yeast uses sugar to make **carbon dioxide**. The carbon dioxide bubbles make the bread rise.

Oxygen    Carbon    Oxygen

**Carbon dioxide molecule**

I like cheese.

I like bread.

Yeast uses sugar to make carbon dioxide bubbles.

Chemical reactions also happen in your body.

When you eat food, your body breaks it down to use for energy, to make bones, and to build muscle.

Chemical reactions happen everywhere!

# How To Say Science Words

**atom**  (AA-tum)

**carbon**   (CAR-buhn)

**chemical reaction**
   (KEH-muh-kuhl  ree-ACK-shun)

**chlorine**   (KLAW-reen)

**electron**  (i-LEK-trahn)

**hydrogen**   (HYE-druh-juhn)

**molecule**   (MAH-lih-kyool)

**nitrogen**   (NYE-truh-jun)

**oxygen**   (OCK-sih-jun)

**particle**   (PAHR-ti-kul)

**phosphorus**   (FAS-fuh-rus)

**sodium**   (SOH-dee-um)

www.ingramcontent.com/pod-product-compliance
Lightning Source LLC
Chambersburg PA
CBHW040148200326
41520CB00028B/7535